iScience
Readers

Living Systems:
Life's Inside Story

by Emily Sohn and Patricia Ohlenroth

Chief Content Consultant
Edward Rock
Associate Executive Director, National Science Teachers Association

NORWOOD HOUSE PRESS
Chicago, Illinois

Norwood House Press
PO Box 316598
Chicago, IL 60631

For information regarding Norwood House Press, please visit our website at
www.norwoodhousepress.com or call 866-565-2900.

Special thanks to: Amanda Jones, Amy Karasick, Alanna Mertens, Terrence Young, Jr.

Editors: Barbara J. Foster, Diane Hinckley
Designer: Daniel M. Greene
Production Management: Victory Productions, Inc.

Library of Congress Cataloging-in-Publication data

Sohn, Emily.

Living systems: life's inside story / by Emily Sohn and Patricia Ohlenroth.
p. cm.

Summary: "Describes the different systems that make up humans and plants:
how and why they both eat, drink, and stay alive. As readers use scientific
inquiry to learn about these specific living systems, an activity based on real
world situations challenges them to apply what they've learned in order to
solve a puzzle"—Provided by publisher.

Includes bibliographical references and index.

ISBN-13: 978-1-59953-427-5 (library edition: alk. paper)
ISBN-10: 1-59953-427-4 (library edition: alk. paper)

1. Biology. 2. Biological systems. I. Ohlenroth, Pat. II. Title.

QH313.S64 2011
570—dc22
2010044737

Manufactured in the United States of America in North Mankato, Minnesota.

165N—012011

CONTENTS

Note to Caregivers:

Throughout this book, many questions are posed to the reader. Some are open-ended and ask what the reader thinks. Discuss these questions with your child and guide him or her in thinking through the possible answers and outcomes. There are also questions posed which have a specific answer. Encourage your child to read through the text to determine the correct answer. Most importantly, encourage answers grounded in reality while also allowing imaginations to soar. Information to help support you as you share the book with your child is provided in the back in the **Additional Notes** section.

Words that are **bolded** are defined in the glossary in the back of the book.

Systems Within Systems

Our world is full of living things. No matter where you live, you see people, other animals, and plants. You might see squirrels in a park or deer in a field, ivy plants in your apartment or spruce trees in the woods, goldfish in a bowl or killer whales in the ocean. These are just a few of the forms life takes on Earth. And inside every one of these living things are many systems that keep life ticking.

Now you're going to learn about these systems that help life thrive. But in Ed's greenhouse, something is not thriving. Is it Ed himself or one of his beloved plants? It is your job to find out.

Which System Isn't Working?

Ed, a family friend, is the new owner of the local greenhouse. He grows plants and flowers that he sells to flower shops and other stores. Ed hasn't been a gardener for long, and he needs help keeping up with the busy summer season. So he has hired you.

Right away you notice that something isn't right at Ed's greenhouse. Four things are going on that seem out of the ordinary. Are they all bad? Are they all really okay? Take a look at the situations and see what you think:

Situation 1:

Ed is huffing and puffing as he works. He's breathing harder than usual. Is something wrong with his heart and lungs?

Situation 2:

Ed is urinating more than usual. Is something wrong with his kidneys or his bladder?

Situation 3:

Ed's favorite potted rose plant is failing. Its roots are coming out of its container. No matter what Ed does—water the plant, fertilize it, give it some rest, give it more sunlight, give it less sunlight—it just gets droopier and droopier.

Situation 4:

Last week Ed's hibiscus plant had large, red, trumpet-shaped flowers. Today the flowers have holes in them. What happened?

To solve this puzzle, you will have to learn about transport systems in humans and in plants. You will have to figure out what happens when the systems are working correctly and what happens when the systems are not working correctly. To do this, you will need to learn about the structure and function of many organs in the human body and the structure and function of many plant parts as well. Read on to get started solving the mystery!

The Blood Bus

You breathe about 20,000 times a day. That's a lot of gas moving into and out of your body. When you inhale, where does the air go in your body? When you exhale, what kind of gas comes out? Is it the same as the gas that goes in?

To move all this gas around, your body has to work like a machine. Blood is a key part of that machine. You may think of blood only when you scrape your knees or get a bloody nose. But blood has the job of carrying the gas from the air you breathe around your body. It also transports nutrients to your body parts. And it carries wastes away. Blood is busy!

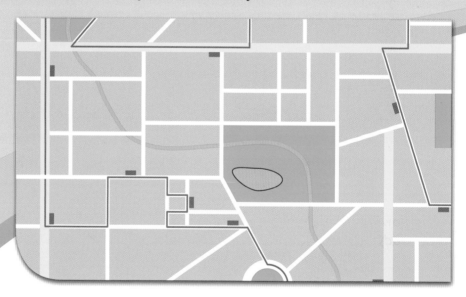

People can move around by bus on a route like this. Nutrients and wastes in your body move around in your blood.

Now think about a bus route in your town or city. Where does the bus go in a day? Does it travel in a straight line and back again? Does it travel in a loop? What does the bus do as it travels? It stops to let passengers get on and off. It stops for fuel. What would happen if the bus didn't make these stops?

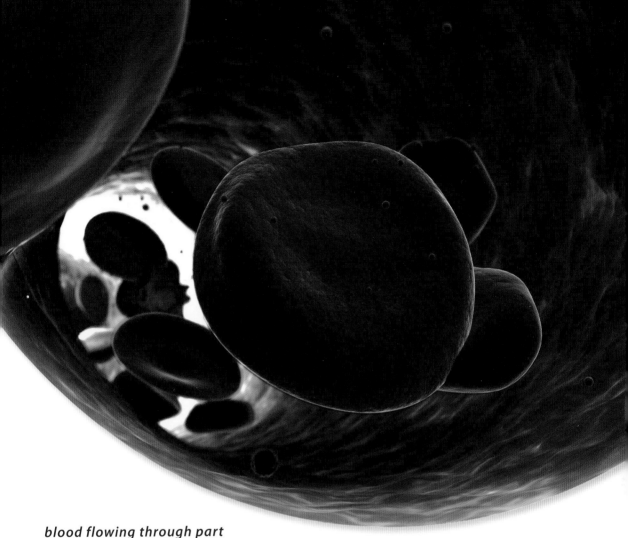

*blood flowing through part
of the body*

Imagine that your body is a city and blood is the bus. But instead of carrying people and objects, the blood carries nutrients and wastes.

Like a bus, the blood in your body follows a route. Try to draw the routes your blood takes through your body. How is the blood's movement like the movement of a bus system? How are the systems different?

Body Basics: Cells

Living things come in all shapes and sizes. They include such plants as trees and grass, and animals, such as people, dogs, and worms. Most living things look very different from you. But you actually share a lot in common with them. For one thing, all living things are made of **cells.** Cells are known as the building blocks of life. They are the basic units of all plants, animals, and other living things. Some **organisms,** such as bacteria, are made of just a single cell. Fish, people, seaweed, and most other living things are made of many cells.

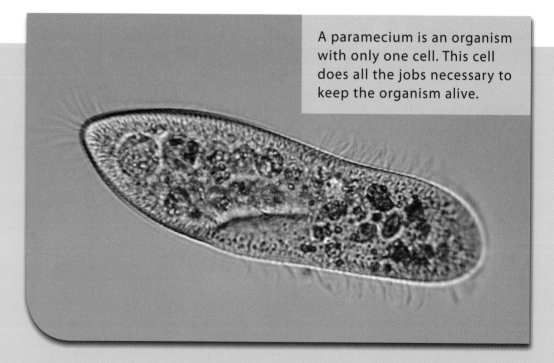

A paramecium is an organism with only one cell. This cell does all the jobs necessary to keep the organism alive.

Cells come in many types. In an organism with many cells, each type of cell has a special job. Think about what you have to do while swimming underwater. You need to kick your legs. You need to move your arms. And you need to think about what you're doing. How might different types of cells help you do all these things?

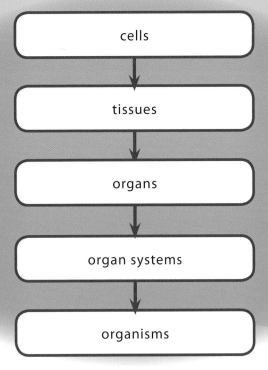

the levels of organization in the body

How Are Cells Organized?

Your body has as many as 100 trillion cells. That's 100,000,000,000,000! All the cells in your body are too small to be seen with the naked eye, but some are bigger than others. They have different shapes, too. A cell's size and shape are related to its job, or function.

Cells are masters of teamwork. In the human body and in many other living things, bunches of cells work together to do the same job. These cells that work together make up **tissues.** Each kind of tissue has specific jobs. Tissues, in turn, make up **organs,** such as the stomach, skin, liver, and lungs.

Groups of organs form **organ systems.** And different systems work together, too.

Blood moves nutrients and wastes around the body. Which organs do you think help blood do its job?

When you play sports, your body uses energy. The food you eat gives you this energy.

How Do Cells Obtain Energy?

You need energy to swim, dive, and play. Every cell in your body needs energy to do its work, too. For example, your brain cells need energy to think. The muscle cells in your legs need energy to kick. Your lung cells need energy to breathe.

The fuel your body needs comes from the food you eat. Your body breaks down food into a few kinds of basic molecules, or particles. Simple sugars are one example. They work like packets of energy for your cells.

a model of the glucose molecule

Blood also brings an important gas called oxygen from the air we breathe to every cell in the body. Cells use oxygen to turn basic food molecules into energy. The process is called **respiration.** Cells need oxygen to work and grow. All living cells get energy through respiration. In cellular respiration, glucose combines with oxygen. (Glucose is a simple sugar made from the elements carbon, hydrogen, and oxygen.) Combining glucose with oxygen gives you carbon dioxide, water, and energy.

glucose + oxygen $\longrightarrow$ carbon dioxide + water + energy

You can read the line above like a sentence: Glucose plus oxygen yields carbon dioxide, water, and energy.

During respiration, cells get rid of some wastes. The waste products of respiration include carbon dioxide gas and water. Your blood carries this stuff away from the cells. Later, your body gets rid of it. Wastes come out when you sweat, exhale, and go to the bathroom. If wastes are not removed from a cell, the cell cannot function properly.

Earlier, you thought about your blood as being like a bus system. What kind of waste do vehicles like buses produce? The waste, or exhaust, of some buses contains a poisonous gas called carbon monoxide as well as other dangerous substances.

What Is the Purpose of Your Body's Transport Systems?

To stay alive, your body needs to move nutrients and wastes around. That movement is so important that your body has a few ways to do it.

As society has many ways to move goods and people around, the body has many ways to move nutrients and wastes around.

The human body has three main systems of transportation. These are the **circulatory system,** the **respiratory system,** and the **excretory system.** The **digestive system** plays a part, too.

Pump It!

The circulatory system begins with the heart. This large pump pushes blood around and around the body. From the heart, blood travels through tubelike vessels. Some of these vessels are quite large; others are tiny. Blood vessels reach every cell in your body. The heart, blood, and blood vessels make up the circulatory system.

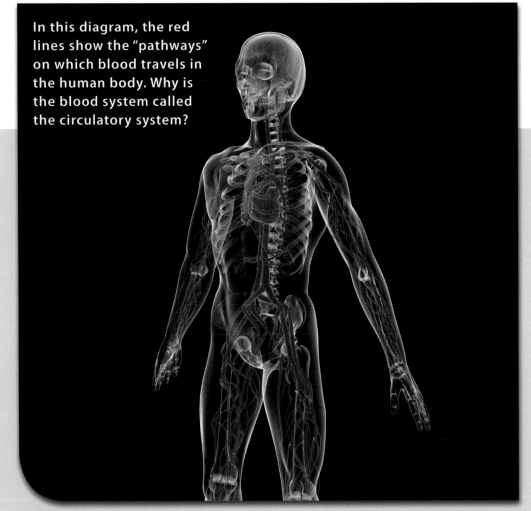

In this diagram, the red lines show the "pathways" on which blood travels in the human body. Why is the blood system called the circulatory system?

What do you think happens to blood when the heart stops pumping? Can you think of any reason why the heart might pump faster than usual?

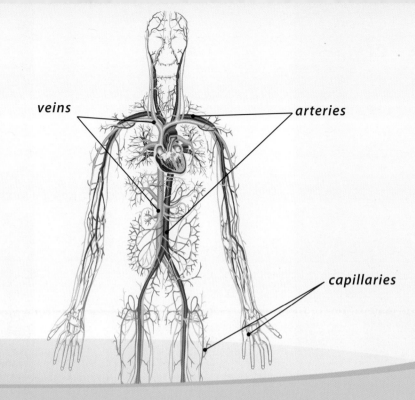

veins   *arteries*

capillaries

You can think of blood vessels as straws of different sizes. Some vessels carry blood away from the heart. These are called **arteries.** In arteries, the blood is full of oxygen. The widest arteries are near the heart. As blood gets farther away from the heart, it moves into smaller vessels.

The smallest blood vessels are called **capillaries.** Their walls are so thin that nutrients and gases can pass right through them into cells. Wastes from the cells can pass through the walls into the blood to be carried away.

When blood flows away from the cells, it is carrying carbon dioxide. The capillaries transport the blood to **veins.** Veins are the vessels that carry blood back to the heart. Veins closer to the heart are bigger than veins farther from the heart.

Blood moves around your body more than 1,000 times a day. Why does blood have to keep circulating?

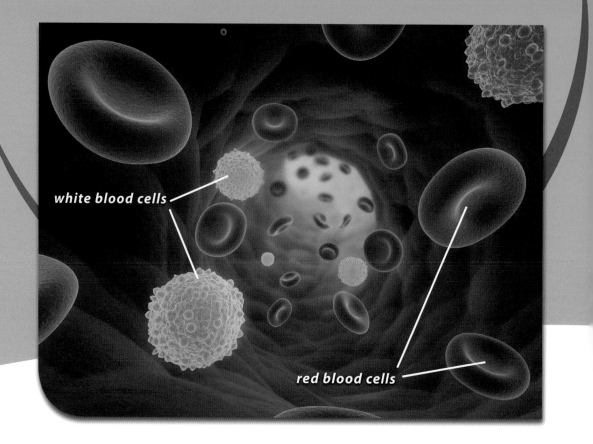

white blood cells

red blood cells

How Do the Blood and the Circulatory System Do Their Work?

Blood cells are the workers that keep nutrients and wastes moving through your body. In each drop of blood, there are millions of blood cells. They come in two types: red blood cells and white blood cells.

Red blood cells are the most common cells in the human body. They move through blood vessels like sticks in a stream. Red blood cells carry oxygen to your cells and carry carbon dioxide away from your cells. What would happen if your red blood cells stopped working very well?

White blood cells are a bit like firefighters. A lot of them show up to "put out the fire."

White blood cells help keep you strong and healthy. They rush to injury sites to start the healing process. Also, they help prevent and fight infections. Have you ever hurt your hand or foot and seen it swell? White blood cells caused the swelling when so many of them showed up in the same place at the same time to help.

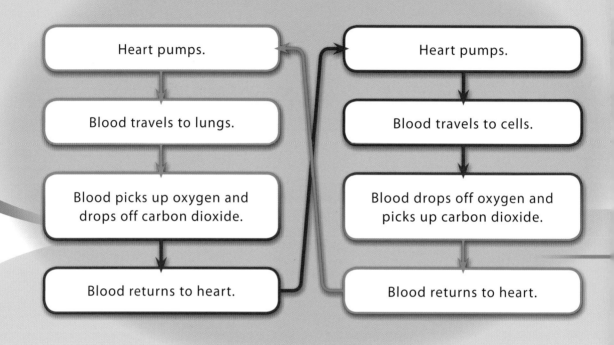

the pathway of blood through the body

Heart pumps.	Heart pumps.
Blood travels to lungs.	Blood travels to cells.
Blood picks up oxygen and drops off carbon dioxide.	Blood drops off oxygen and picks up carbon dioxide.
Blood returns to heart.	Blood returns to heart.

A bus follows the same route over and over. In the same way, blood travels through your body in a jagged loop. First, your heart pumps blood to your lungs. There, it picks up oxygen that comes in when you inhale. This oxygen-rich blood flows back to your heart. The heart then pumps it out to all the cells of your body.

At the cells, oxygen and nutrients pass through capillaries directly into the cells. The cells use these deliveries for respiration. That gives them energy to do their jobs.

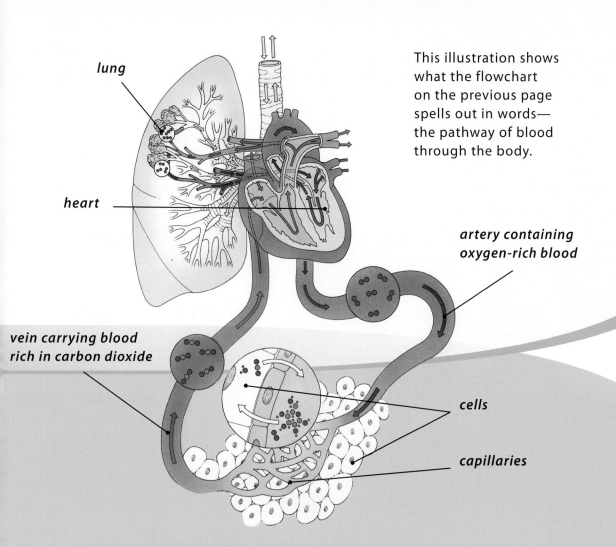

lung

heart

artery containing
oxygen-rich blood

vein carrying blood
rich in carbon dioxide

cells

capillaries

This illustration shows
what the flowchart
on the previous page
spells out in words—
the pathway of blood
through the body.

When carrying out their jobs, cells produce wastes, such as carbon dioxide. These wastes flow through the cell walls into the bloodstream. Veins deliver the waste-filled blood to tiny air sacs in the lungs. There, the carbon dioxide is released, and you breathe it out. When you breathe in again, the blood picks up more oxygen. This blood then flows back to the heart, and the cycle starts all over again.

How might your body respond if you are working really hard? Think back to Ed in his greenhouse. Is there a good reason why he might be huffing and puffing? Or is something wrong?

CONNECTING TO HISTORY

Discovery of the Pumping Heart

Today, just about everyone knows that the heart pumps blood through the body. Nobody knew that in 1578. That year, a boy named William Harvey was born in England. When Harvey was little, people thought the liver turned food into blood. (The liver is another organ in your body.) They thought the body used blood as fuel.

Harvey became a medical doctor. He started poking around the human body.

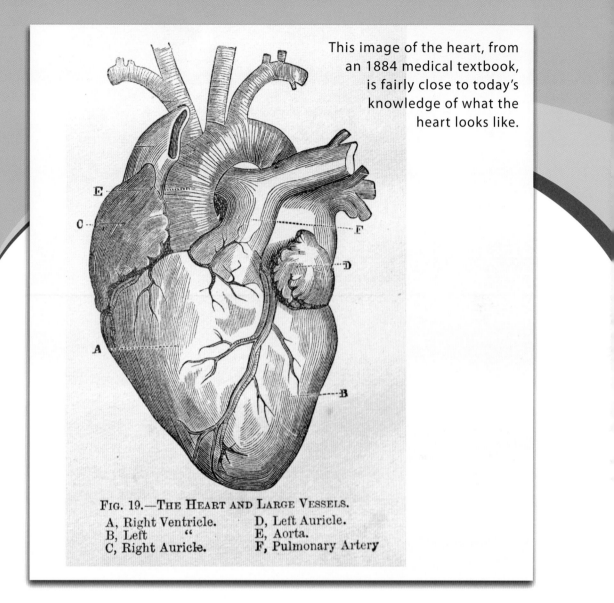

This image of the heart, from an 1884 medical textbook, is fairly close to today's knowledge of what the heart looks like.

FIG. 19.—THE HEART AND LARGE VESSELS.

A, Right Ventricle.   D, Left Auricle.
B, Left   "   E, Aorta.
C, Right Auricle.   F, Pulmonary Artery

Harvey looked inside the dead bodies of people and animals. At the time, people found that unsettling. But Harvey learned a lot about the heart and blood vessels that way. He was interested in how blood flowed through the body.

Based on his observations, Harvey published *An Anatomical Study of the Motion of the Heart and of the Blood in Animals* in 1628. He became the first person to accurately describe how the heart pumps blood around the body.

Maybe *your* blood will help save someone's life someday!

SCIENCE AT WORK

Blood Bank Technician

When people get hurt badly, they can lose a lot of blood. To survive, they may need blood from other people. It is the job of a blood bank technician to collect blood from volunteers. This blood gets sent to hospitals.

At a blood bank, a technician puts a needle right into a volunteer's vein. The person's blood flows into a collection tube. To perform this job, a technician must know where the veins are. He or she must also know how to pull the needle out and stop the flow of blood.

People can pass out if there is not enough oxygen in their brains. Why should people stand up slowly after giving blood?

The Breath of Life

Oxygen is one of the pillars of life. You've learned that cells need it. You also know that the circulatory system delivers it. None of this could happen without the respiratory system. This is the system that helps you breathe. It includes the lungs, certain muscles, passageways for air, and blood vessels. That's right! The same blood vessels you find in the circulatory system are part of the respiratory system. Sometimes a body part belongs to more than one system.

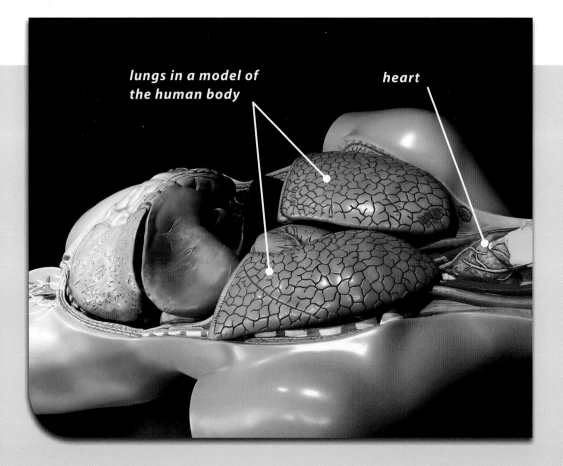

lungs in a model of the human body

heart

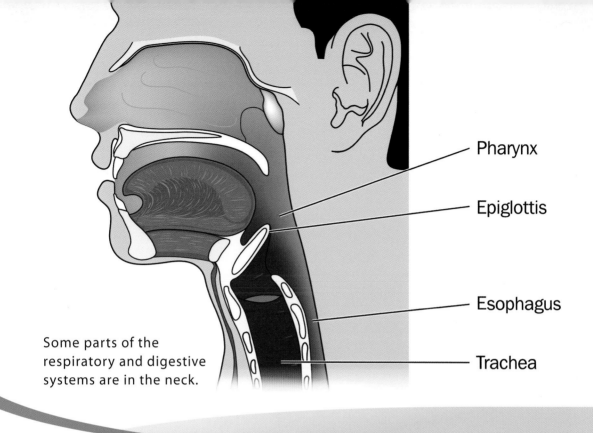

Pharynx

Epiglottis

Esophagus

Trachea

Some parts of the respiratory and digestive systems are in the neck.

When you breathe, air goes into your mouth or nose and down through your **pharynx,** which is a pipelike organ. Food travels down the same pipe. At the bottom of the pipe, food goes one way (through the esophagus) and air goes another. A flap called the **epiglottis** closes when we swallow. This keeps food and liquid out of the trachea. The **trachea** is the pipe that carries air past the epiglottis and to the lungs. You may have heard it called the windpipe.

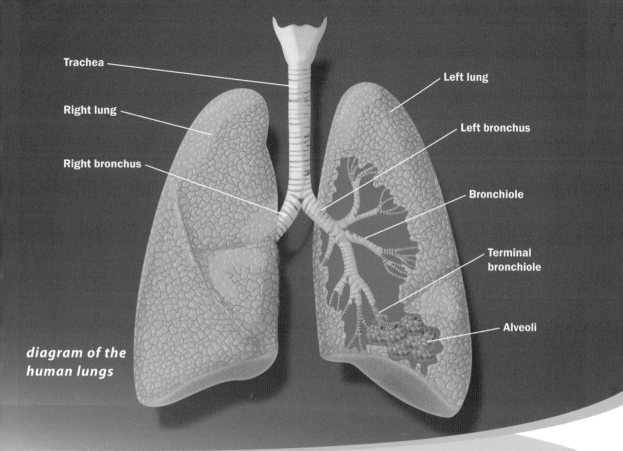

Trachea

Right lung

Right bronchus

Left lung

Left bronchus

Bronchiole

Terminal bronchiole

Alveoli

diagram of the human lungs

In the lungs, the trachea branches off into smaller and smaller pipes. These airways are called **bronchioles.** At the end of the bronchioles are tiny air sacs called **alveoli.** A net of blood vessels covers each of these sacs. At the alveoli, blood vessels pick up oxygen out of the air you breathe in. And they release carbon dioxide into the lungs, from where it gets breathed out of the body.

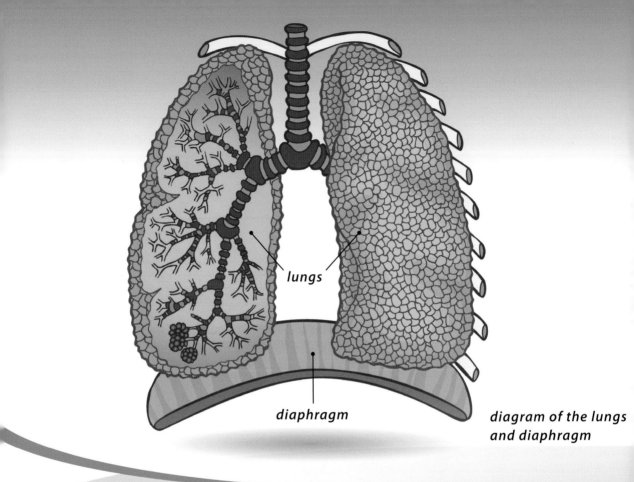

lungs

diaphragm

diagram of the lungs and diaphragm

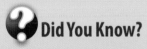 **Did You Know?**

Hiccups happen when the diaphragm quickly contracts over and over. As air rushes in, the epiglottis closes. That makes you go "hic"!

Muscles help air move through your body, too. A muscle called the diaphragm sits beneath the lungs. When you breathe in, the diaphragm contracts, like a fist. This gives the lungs more room to expand and fill with air. When you breathe out, the diaphragm loosens up. This helps push carbon dioxide out of the lungs.

A greenhouse is a very busy place in the summertime!

How do you think your body reacts if your cells aren't getting enough oxygen? Think back to your iScience puzzle. Is there any situation that makes you think of the respiratory system? Remember, Ed and his employees—including you—are working very hard. You are probably lifting heavy plants and working fast because the greenhouse business is so busy in the summer. Would that affect Ed's body in any way?

Everything your body does requires fuel. Much of that fuel comes from the food you eat. The digestive system breaks food down into tiny parts. It also takes nutrients, like vitamins, minerals, and proteins, out of foods so your body can use them.

Your digestive tract is about as long as this school bus!

Your teeth, tongue, and digestive juices in saliva begin breaking down food in your mouth. After you swallow, muscles push food through the digestive tract. This passageway runs from your mouth to the part of your body where solid wastes come out. The tract is a tube that measures about 30 feet (9 meters) long.

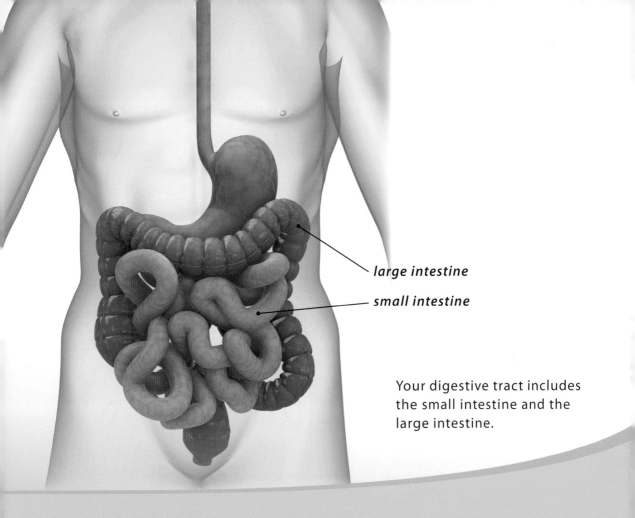

large intestine

small intestine

Your digestive tract includes the small intestine and the large intestine.

From your mouth, food travels through the esophagus to your stomach, where it gets mushier and mushier. To help out, the liver and pancreas produce juices that help break food down. Next, the mush continues to the **small intestine.** There, nutrients pass to the blood stream, which carries them around the body. Your body can't use all the parts of your food. The leftover stuff passes to the **large intestine.**

How Does Material Move Through the Excretory System?

No matter how healthy you are, your body makes a lot of waste. The excretory system is like your body's garbage truck. It takes the wastes away.

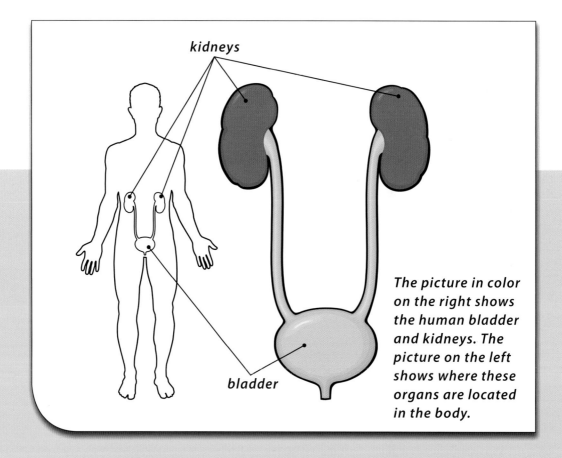

kidneys

bladder

The picture in color on the right shows the human bladder and kidneys. The picture on the left shows where these organs are located in the body.

First, working cells put out water and chemicals as waste products. These and other leftover products enter the blood. They travel to the **kidneys,** which filter out much of the waste. The kidneys also remove extra salt and water from the blood. Most people have two kidneys. Together, they filter about 200 quarts of blood a day. The cleaned blood continues to circulate in the body. Water and other wastes become urine.

When the bladder is full, the urge to urinate is strong.

RESTROOMS →

The body stores urine, a liquid, in the **bladder** until you urinate by loosening certain muscles. Then, the urine comes out through a tube into the toilet. Fiber and other undigested food parts collect in the large intestine. When you defecate, they pass out of the body in a more solid form.

Think back to the iScience puzzle. Ed is working hard in a hot greenhouse in the summer. What might cause him to urinate more than usual? Is that normal or is something wrong with Ed?

Transport Systems in Plants

As you've seen, it takes a lot of work to keep animals, including people, alive. Lots of systems are involved. Like animals, plants have transport systems that move nutrients and wastes. And like in animals, these systems are made of various tissues. The tissues are made of cells.

Leaves are part of a tree's transport system.

Just as human transport systems involve many structures in the body, plant transport systems involve plant structures such as roots, stem, and leaves.

leaves

stem

This picture shows a plant's stem, leaves, and roots.

roots

Plants don't have arms, skin, or brains. But their tissues have jobs much like animal tissues do. That's because plants need food and water, too. They also need to get rid of wastes.

Plants have just three basic types of tissue groups. Most of a land plant is made of ground tissue. This tissue makes and stores nutrients. It also supports a plant just as beams support a building. What supports the human body?

A waxy layer called dermal tissue covers the outside of leafy plants. This waxy layer protects the plant. It keeps harmful invaders out. And it helps keep water in. Which part of your body does these things for you?

The third type is vascular tissue. This is what moves stuff around inside a plant. It carries water, minerals, and nutrients. How do systems in your body compare to plant systems?

Xylem hardens and forms one ring each year. By counting the rings, you can tell how old the tree was when it was cut down.

tree rings made of hardened xylem

Have you ever hugged a tree this big?

? Did You Know?

The giant sequoia is the world's largest type of tree. It can grow higher than a building with 25 floors. That means a sequoia is able to raise water to a height of more than 250 feet (76 meters) above the ground! Why do you think the top of a tree needs water?

To get things moving, the vascular system uses two kinds of tissues. These are called **xylem** and **phloem.** Xylem carries water and minerals from the roots up to the rest of the plant. In trees, xylem hardens to make rings of wood in the trunk. One new ring forms each year. Phloem moves food down from the leaves to the rest of the plant.

What happens if transportation routes get blocked?

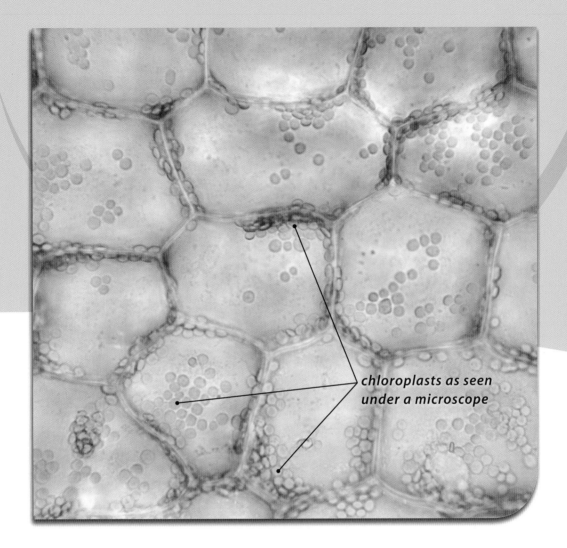

chloroplasts as seen under a microscope

How Does a Plant Get Food?

Plants need food, but they don't eat the way animals do. Instead, they make their own food, in a process called **photosynthesis.** The basic ingredients are simple. All they need are sunlight, carbon dioxide from air, and water.

In plants, most photosynthesis happens in the leaves. Plant cells have special parts called **chloroplasts.** These parts are like cooks. They make the food.

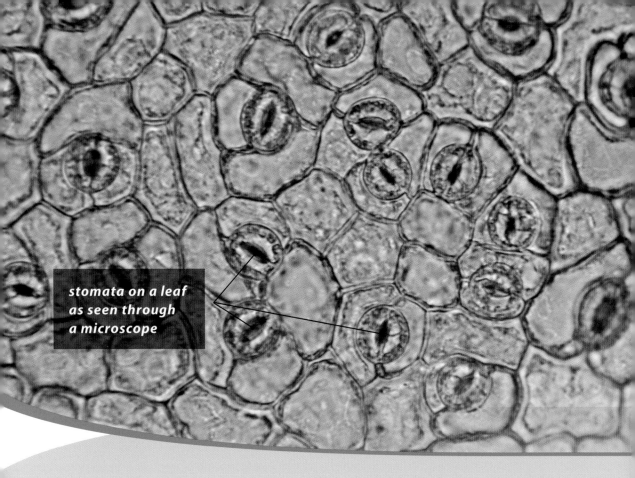

stomata on a leaf
as seen through
a microscope

Chloroplasts use energy from the Sun, along with water and carbon dioxide, to make sugars. Plant roots take in water from the soil, and then the water moves through the stem and into the leaves. The carbon dioxide comes in through openings on the leaves called **stomata.** Stomata can open and close to allow gases in and out, just as gases enter and leave your body through your mouth and nose.

Photosynthesis makes sugars, which the plant uses to grow. The process also makes oxygen. Plants get rid of this gas through their stomata. When you inhale, you take in oxygen that came from plants. When you exhale, you breathe out carbon dioxide that plants take in.

Algae are like plants, but they lack true stems, roots, and leaves.

Plants aren't the only organisms that perform photosynthesis. Plantlike organisms called algae do, too. In the ocean, tiny algae generate half of the oxygen on Earth.

❓ Did You Know?

Often, the presence of algae means that a body of water has an imbalance of nutrients. Many algae grow well in the presence of nitrogen and phosphorus. Over time, algae can prevent sunlight from penetrating the water, which means other plants will die off. Without plants giving off oxygen into the water, fish and other organisms die off as well.

Transpiration from plants helps keep the jungle a humid place.

Thirsty Plants

Plants need more than just food to live. They also need water. On land, roots absorb water mixed with minerals from the soil. From there, water and minerals travel through the xylem to the leaves. When guard cells in the leaves open, they let extra wastewater out as a gas. The gas is called vapor. This process is called **transpiration.**

What would happen to a plant that couldn't get enough water through its roots? Think back to Ed and the greenhouse. Could not getting enough water explain why the mini potted rose plant is not thriving?

Transpiration helps water move up through a plant similar to the way a beverage moves up through a straw.

For a land plant, losing water is like sweating. The guard cells close to help keep a plant from losing too much water. If a plant loses too much water, it will dry out and die.

Transpiration also allows a thirsty plant to get more water and nutrients from the soil. It's like drinking out of a straw. When water vapor passes out of the top of a plant, more water gets sucked up at the roots. The water brings nutrients with it.

What happens if animals eat the leaves of a plant? What happens if animals eat the stems?

Think again about the greenhouse. Could animals explain the condition of any of the plants there?

In the puzzle, you were given four situations to look at. You realized that something was not right in Ed's greenhouse. Let's reconsider the situations you observed. Below, each situation includes an educated guess, or hypothesis, about what could be going on. For each hypothesis, consider whether it is a sign that something—or someone— is not thriving.

Situation 1:

Ed is huffing and puffing as he works. He's breathing harder than usual. But that's normal when someone works hard lifting heavy objects like plants. Ed's heart and lungs are probably just fine.

Situation 2:

Ed is urinating more than usual. But in the summer in a hot greenhouse he is probably drinking a lot of water to replace what he loses by sweating. Some of this extra water will naturally come out as urine. Ed's kidneys and bladder are also probably fine.

Situation 3:

Ed's favorite mini potted rose plant is failing. Its roots are coming out of its container. With the roots coming out of the container, they are not in the soil where the water and nutrients are. Without water and nutrients, the plant isn't getting what it needs to thrive. Ed should try planting the mini rose in a bigger pot.

Situation 4:

Ed's hibiscus flowers suddenly don't look right. When they first bloomed, the flowers looked fine. But now they have holes in them. Perhaps animals have been nibbling at them, thus explaining their looks.

These are just some of the possible answers to these mysteries. Do you have any other hypotheses of your own to add?

In this book, you learned about how living things grow and thrive. You learned that animals depend on plants and algae. Through photosynthesis, these green organisms supply us with oxygen, which we need to live. Think about the ways that human activity sometimes affects plant life.

In April 2010, an oil rig exploded in the Gulf of Mexico. For five months, millions of barrels of oil gushed out of a hole in the bottom of the sea. How do you think the spill affected plants? How did it affect animals?

Some marsh grasses near the Gulf were covered with oil from the spill. Draw a model of how a plant gets food. Show how oil will affect the plant. In turn, how will it affect the animals that eat those plants?

There are probably more living things on Earth than you can count. No matter how different from each other they may seem, they all need certain things to survive, thrive, and stay alive.

Oil from the 2010 Gulf of Mexico spill threatened the health of these grasses in Mississippi.

GLOSSARY

alveoli: tiny air sacs in the lungs where oxygen is picked up and carbon dioxide is released.

arteries: blood vessels that carry blood with oxygen away from the heart.

bladder: place in the body where urine is stored.

bronchioles: small airways in the lungs.

capillaries: smallest blood vessels in the human body.

cells: the basic units of all living things.

chloroplasts: special plant cells that absorb energy from the Sun.

circulatory system: organs and tissues that work together to transport nutrients and wastes to and from all the cells in the human body.

digestive system: organs and tissues that work together to break down food the body can use for energy.

epiglottis: flap that closes the pipe to the lungs when food or drink is swallowed.

excretory system: organs and tissues that work together to remove waste from the body.

kidneys: organs that filter waste, salt, and water from the blood.

large intestine: part of the digestive system where unused food parts collect and pass through.

organisms: living things.

organ systems: groups of organs in the human body that work together.

organs: structures of special tissue that have a special job in the body.

pharynx: passageway for food and air in the throat.

phloem: part of the plant that transports food.

photosynthesis: process through which plants make their own food.

respiration: process that all living cells use to make energy.

respiratory system: organs and tissues that work together to transport oxygen and carbon dioxide to and from all the cells in the body.

small intestine: part of the digestive system where food is absorbed and passed into the bloodstream.

stomata: openings on a plant that allow gases to pass in and out.

tissues: cells with the same structure that work together.

trachea: the pipe that carries air past the epiglottis and to the lungs.

transpiration: process through which plants absorb and release water.

veins: blood vessels that carry blood with carbon dioxide to the heart.

xylem: cells that transport water and minerals around a plant, and that harden in trees to form woody rings.

FURTHER READING

Mary K Corcoran. *The Quest to Digest* Watertown, MA: Charlesbridge Publishing, 2006.

biology4kids.com. **Photosynthesis.**
http://www.biology4kids.com/files/plants_photosynthesis.html

The Franklin Institute. **The Human Heart.**
http://www.fi.edu/learn/heart/index.html

University of Southern California Sea Grant Program. Monterey. Science Kids. **Science Games for Kids.**
http://www.sciencekids.co.nz/gamesactivities/plantsgrow.html

ADDITIONAL NOTES

The page references below provide answers to questions asked throughout the book. Questions whose answers will vary are not addressed.

Page 8: If the bus didn't stop, people wouldn't get where they're going.

Page 9: Like a bus follows a route through town, blood follows a route through the body. Both blood and a bus pick up fuel and let out wastes. But the bus stops to do its business; blood is always moving.

Page 11: The heart and lungs.

Page 14: Buses can produce exhaust fumes, including carbon dioxide.

Page 16: When the heart stops pumping, blood doesn't flow to the brain and this could cause death. The heart might beat faster when a person exercises or is frightened.

Page 17: Blood keeps circulating so that nutrients and oxygen continue flowing and that waste doesn't build up.

Page 18: If your red blood cells stopped working well, the rest of your cells wouldn't get the fuel they need or be able to get rid of waste.

Page 24: After people give blood, there is less blood in their body, so there is also temporarily less oxygen in their body, too. There might be less oxygen getting to their brains and so they may feel dizzy.

Page 35: The skeleton supports the human body. The circulatory, respiratory, digestive, and excretory systems in the human body move nutrients and wastes around your body just like plant systems move nutrients and wastes around a plant's body.

Page 36: If transportation routes get blocked, the tree will not get the nutrients it needs.

Page 40: A plant that couldn't get enough water through its roots would die.